Modeling and solving linear programming with R

Jose M Sallan   Oriol Lordan   Vicenc Fernandez

Modeling and Solving Linear Programming with R

Authors:

Jose M. Sallan, Oriol Lordan, Vicenc Fernandez

Universitat Politècnica de Catalunya

ISBN: 978-84-944229-3-5

DOI: http://dx.doi.org/10.3926/oss.20

Contents

CHAPTER 1

Introduction

This book is about using linear programming to help making better decisions in the organizational context. Linear programming is one of the most useful and extensively used techniques of operational research. It is one special case of mathematical optimization, where the function to optimize and the constraints are linear functions of the decision variables. Posterior developments of linear programming include the possibility of defining some decision variables as integer, widening the range of problems solvable by linear programming considerably.

This is the first of a series of books that act as a support of a pedagogical program based on teaching operational research techniques with R. R [6] is a programming language and software environment for statistical computing and graphics. The R language is widely used among statisticians and data miners for developing statistical software and data analysis. It is an open source programming environment, that runs in most operating systems. The strength of R comes from the large number of libraries developed by a lively community of software developers. Within the context of this teaching program, the objective of this book is twofold. On the one side, our aim is to present a pragmatic introduction to linear programming, presenting through practical examples the possibilities of modeling through linear programming situations of decision making in the organizational context. On the other side, some libraries to solve linear programming models are presented, such as Rglpk [7], lpSolve [1] and Rsymphony [3].

To achieve these aims, the book is organized as follows. In 2.6.2 are developed the basics of linear programming: an introduction of formulation of linear models, an introduction to the features of the optimum of a linear program, including duality analysis, and to the formulation and solution of linear programs including integer variables. The chapter concludes with an introduction to the use of linear programming solvers in R.

chapter 3 includes ten optimization problems solvable by linear programming. Each of the problems is presented with the following structure: after presenting the problem, a solution through linear programming is offered. Then we show how to solve the problem in R. There are several ways to parse a problem into a R solver. In this collection of problems, we show how to use a standard linear programming syntax, such as CPLEX, and how to enter the model using the R syntax.

We have chosen to use online resources to keep this book updated. In http://bit.ly/1zkJpVw we are keeping a list of linear programming solvers, together with its implementation in R. We encourage readers to send us a comment if they find the information incomplete or not updated. All the source code used in this book is stored and updated in the https://github.com/jmsallan/linearprogramming GitHub repository.

We hope that this book becomes a valuable resource to everybody interested in a hands-on introduction to linear programming, that helps to reduce the steep of the learning curve to implement code including resolution of linear programming models.

CHAPTER 2

Solving linear programming

2.1 An introduction to linear programming

Linear programming is one of the most extensively used techniques in the toolbox of quantitative methods of optimization. Its origins date as early as 1937, when Leonid Kantorovich published his paper *A new method of solving some classes of extremal problems*. Kantorovich developed linear programming as a technique for planning expenditures and returns in order to optimize costs to the army and increase losses to the enemy. The method was kept secret until 1947, when George B. Dantzig published the simplex method for solving linear programming [2]. In this same year, John von Neumann developed the theory of duality in the context of mathematical analysis of game theory.

One of the reasons for the popularity of linear programming is that it allows to model a large variety of situations with a simple framework.

Furthermore, a linear program is relatively easy to solve. The simplex method allows to solve most linear programs efficiently, and the Karmarkar interior-point methods allows a more efficient solving of some kinds of linear programming.

The power of linear programming was greatly enhanced when came the opportunity of solving integer and mixed integer linear programming. In these models all or some of the decision variables are integer, respectively. This field was opened by the introduction of the branch and bound method by Land and Doig. Later other algorithms have appear, like the cutting plane method. These techniques, and the extension of computing availability, have increased largely the possibilities of linear programming.

In this chapter we will provide a brief introduction to linear programming, together with some simple formulations. We will also provide an introduction to free software to solve linear programming in R, in particular the R implementations of lp_solve and GLPK through the libraries lpSolve, Rglpk and Rsymphony, among others. chapter 3 introduces some applications of linear programming, through a collection of solved linear programming problems. For each problem a posible solution through linear programming is introduced, together with the code to solve it with a computer and its numerical solution.

2.2 Linear programming formulation

2.2.1 The structure of a linear program model

Roughly speaking, the linear programming problem consists in *optimizing* (that is, either minimize or maximize) the value of a linear *objective function* of a vector of *decision variables*, considering that the variables can only take the values defined by a set of linear *constraints*. Linear programming is a case of *mathematical programming*, where objective function and constraints are linear.

A formulation of a linear program in its canonical form of maximum is:

$$\text{MAX } z = c_1 x_1 + c_2 x_2 + \cdots + c_n x_n$$
$$\text{s. t. } a_{11} x_1 + a_{12} x_2 + \cdots + a_{1n} x_n \leq b_1$$
$$a_{21} x_1 + a_{22} x_2 + \cdots + a_{2n} x_n \leq b_2$$
$$\cdots$$
$$a_{m1} x_1 + a_{m2} x_2 + \cdots + a_{mn} x_n \leq b_m$$
$$x_i \geq 0$$

The model has the following elements:

- An *objective function* of the n decision variables x_j. Decision variables are affected by the *cost coefficients* c_j

- A set of m *constraints*, in which a linear combination of the variables affected by coefficients a_{ij} has to be less or equal than its *right hand side value* b_i (constraints with signs greater or equal or equalities are also possible)

- The bounds of the decision variables. In this case, all decision variables have to be nonnegative.

The constraints of the LP define the *feasible region*, which is the set of values that satisfy all constants. For a LP of n variables, the feasible region is a n-dimensional convex polytope. For instance, for $n = 2$ the feasible region is a convex polygon.

The LP formulation shown above can be expressed in matrix form as follows (cap bold letters are matrices and cap small bold letters are column vectors):

$$\text{MAX } z = \mathbf{c}'\mathbf{x}$$

$$\text{s. t. } \mathbf{Ax} \leq \mathbf{b}$$

$$\mathbf{x} \geq 0$$

Using the same matrix syntax, we can write the *canonical form* of minimum of a linear program as:

$$\text{MIN } z = \mathbf{c}'\mathbf{x}$$

$$\text{s. t. } \mathbf{Ax} \geq \mathbf{b}$$

$$\mathbf{x} \geq 0$$

Another usual way to express a linear program is the *standard form*. This form is required to apply the simplex method to solve a linear program. Here we have used OPT to express that this form can be defined for maximum or minimum models.

$$\text{OPT } z = \mathbf{c}'\mathbf{x}$$

$$\text{s. t. } \mathbf{Ax} = \mathbf{b}$$

$$\mathbf{x} \geq 0$$

An additional condition to use the simplex method is that righthand side values $\mathbf{b} \geq 0$. All other parameters are not restricted in sign.

2.2.2 A simple example of a PL model

Let's consider the following situation:

A small business sells two products, named Product 1 and Product 2. Each tonne of Product 1 consumes 30 working hours, and each tonne of Product 2 consumes 20 working hours. The business has a maximum of 2,700 working hours for the period considered. As for machine hours, each tonne of Products 1 and 2 consumes 5 and 10 machine hours, respectively. There are 850 machine hours available.

Each tonne of Product 1 yields 20 M€ of profit, while Product 2 yields 60 M€ for each tonne sold. For technical reasons, the firm must produce a minimum of 95 tonnes in total between both products. We need to know how many tonnes of Product 1 and 2 must be produced to maximize total profit.

This situation is apt to be modeled as a PL model. First, we need to define the *decision variables*. In this case we have:

- $P1$ number of tonnes produced and sold of Product 1

- $P2$ number of tonnes produced and sold of Product 2

The *cost coefficients* of these variables are 20 and 60, respectively. Therefore, the *objective function* is defined multiplying each variable by its corresponding cost coefficient.

The constraints of this LP are:

- A constraint WH making that the total amount of working hours used in Product 1 and Product 2, which equals $30P1 + 20P2$, is less or equal than 2,700 hours.

- A similar constraint MH making that the total machine hours $5P1 + 10P2$ are less or equal than 850.

- A PM constraint making that the total units produced and sold $P1 + P2$ are greater or equal than 95.

Putting all this together, and considering that the decision variables are nonnegative, the LP that maximizes profit is:

$$\text{MAX } z = 20P1 + 60P2$$
$$\text{s.t. WH) } 30P1 + 20P2 \leq 2700$$
$$\text{MH } 5P1 + 10P2 \leq 850$$
$$\text{PM) } P1 + P2 \geq 95$$
$$P1 \geq 0, \ P2 \geq 0$$

2.2.3 A transportation problem

Let's consider a *transportation problem* of two origins a and b, and three destinations 1, 2 and 3. In Table 2.1 are presented the cost c_{ij} of transporting one unit from the origin i to destination j, and the maximal capacity of the origins and the required demand in the destinations. We need to know how we must cover the demand of the destinations at a minimal cost.

	1	2	3	capacity
a	8	6	3	70
b	2	4	9	40
demand	40	35	25	

Table 2.1: Parameters of the transportation problem

This situation can be modeled with a LP with the following elements:

- Decision variables of the form x_{ij}, representing units transported from origin i to destination j

- An objective function with cost coefficients equal to c_{ij}

- Two sets of constraints: a less or equal set of constraints for each origin, limiting the units to be transported, and a greater of equal set of constraints representing that the demand of each destination must be covered.

The resulting LP is:

$$\text{MIN } z = 8x_{a1} + 6x_{a2} + 3x_{a3} + 2x_{b1} + 4x_{b2} + 9x_{b3}$$
$$\text{s.a. ca) } x_{a1} + x_{a2} + x_{a3} \leq 70$$
$$\text{cb) } x_{b1} + x_{b2} + x_{b3} \leq 40$$
$$\text{d1) } x_{a1} + x_{b1} \geq 40$$
$$\text{d2) } x_{a2} + x_{b2} \geq 35$$
$$\text{d3) } x_{a3} + x_{b3} \geq 25$$
$$x_{ij} \geq 0$$

2.2.4 Transformations of elements of a LP

Transforming the objective function of a linear program is straightforward. A MAX problem can be transformed into MIN (and vice versa) changing the sign of the cost coefficients:

$$\text{MIN } z = \mathbf{c}'\mathbf{x} \Leftrightarrow \text{MAX } z' = -\mathbf{c}'\mathbf{x}$$

Nonequality constraints can be transformed changing the signs of all terms of the constraint:

$$a_{i1}x_1 + \cdots + a_{in}x_n \leq b_i \Leftrightarrow -a_{i1}x_1 - \cdots - a_{in}x_n \geq -b_i$$

A nonequality constraint can be turned into equality by adding nonnegative variables:

$$a_{i1}x_1 + \cdots + a_{in}x_n \leq b_i \Rightarrow a_{i1}x_1 + \cdots + a_{in}x_n + s_i = b_i$$
$$a_{k1}x_1 + \cdots + a_{kn}x_n \geq b_k \Rightarrow a_{k1}x_1 + \cdots + a_{kn}x_n - e_k = b_k$$
$$s_i \geq 0, \ e_k \geq 0$$

Less than equal constraints are turned into equality by adding slack variables s_i, and greater than equal constraints by excess variables e_k. If the original constraints have to be maintained, both types of variables have to be nonnegative.

Finally, decision variables can also be transformed. A nonpositive variable x_i can be replaced by a nonnegative variable x_i' making $x_i' = -x_i$. A variable unconstrained in sign x_k can be replaced by two nonnegative variables x_k', x_k'' by making $x_k = x_k' - x_k''$.

2.2.5 Turning a PL into standard form

A usual transformation of a PL model is turning all constraints into equalities adding slack and excess variables. This is required to solve the PL using any version of the simplex algorithm. For instance, the model defined in subsection 2.2.2 can be put into standard form making:

$$\text{MAX } z = 20P1 + 60P2$$
$$\text{s.t. WH) } 30P1 + 20P2 + h_W = 2700$$
$$\text{MH } 5P1 + 10P2 + h_M = 850$$
$$\text{PM) } P1 + P2 - e_P = 95$$
$$P1, \ P2, \ h_W, h_M, \ e_P \geq 0$$

where h_W and h_M are equal to the working and machine hours, respectively, not used in the proposed solution, and e_P equals the total production made over the minimal value required of 95. Note than slack and excess variables have to be also nonnegative.

In the standard form, any constraint that was an inequality in the original form will have its corresponding slack or excess variable equal to zero when it is satisfied with the equal sign. Then we will say that this constraint is *active*. If its corresponding slack or excess variable holds with the inequality sign, its corresponding variable will be positive, and the constraint will be not active.

2.3 Solving the LP

The most extended procedure to solve the LP is the *simplex algorithm*, developed by George Bernard Dantzig in 1947. This method takes advantage of the fact that the optimum or optima of a LP can be found exploring its basic solutions. A *basic solution* of a LP in standard form of n variables and m constraints has the following properties:

- has $n - m$ *nonbasic variables* equal to zero: $\mathbf{x}_N = 0$

- has m *basic variables* greater or equal to zero: $\mathbf{x}_N \geq 0$

When one or more basic variables equal zero, the solution is called *degenerate*. The basic solutions correspond to the *vertices of the feasible region*.

The strategy of the simplex method consists in:

- Finding an initial basic solution

- Explore the basic solutions moving in the direction of maximum local increase (MAX) or decrease (MIN) of the objective function

- Stop when an optimal solution is found

The software that solves LPs uses usually the simplex algorithm, or the *revised simplex algorithm*, a variant of the original simplex algorithm that is implemented more efficiently on computers. Other algorithms exist for particular LP problems, such as the transportation or transshipment problem, or the maximum flow problem.

Another approach to solve LPs is the *interior point algorithm*, developed by Narenda Karmarkar [4]. This algorithm has been proven as particularly useful in large problems with sparse matrices. Contrarily to the simplex approach, this algorithm starts from a point inside the feasible region, and approaches the optimum iteratively.

2.4   Duality in linear programming

Let's consider a MAX linear program in its canonical form:

$$\text{MAX } z = \mathbf{c}'\mathbf{x}$$

$$\text{s. t. } \mathbf{A}\mathbf{x} \leq \mathbf{b}$$

$$\mathbf{x} \geq 0$$

The following linear program, expressed in MIN canonical form, is the *dual* of the program above, called the *primal*:

$$\text{MIN } w = \mathbf{u}'\mathbf{b}$$

$$\text{s. t. } \mathbf{u}'\mathbf{A} \geq \mathbf{c}'$$

$$\mathbf{u} \geq 0$$

Note that each variable of the dual is linked with a constraint of the primal, since both share the same b_j parameter. Accordingly, each constraint of the dual is linked to a variable of the primal, as both share the same c_i parameter.

If the linear program is not expressed in canonical form, it can be turn into canonical form using the transformations defined in section 2.2. More conveniently, the dual can be obtained applying the transformations defined in Table 2.2 for the original formulation of the model.

MAX	MIN
constraint $\leq$	variable ≥ 0
constraint $\geq$	variable ≤ 0
constraint $=$	variable unconstrained
variable ≥ 0	constraint $\geq$
variable ≤ 0	constraint $\leq$
variable unconstrained	constraint $=$

Table 2.2: Primal to dual conversion table

2.4.1 Obtaining the dual of the LP

Let's consider the LP formulated in subsection 2.2.2:

$$\text{MAX } z = 20P1 + 60P2$$

$$\text{s.t. WH)} \ 30P1 + 20P2 \leq 2700$$

$$\text{MH } 5P1 + 10P2 \leq 850$$

$$\text{PM)} \ P1 + P2 \geq 95$$

$$P1 \geq 0, \ P2 \geq 0$$

The dual of this model will have three decision variables, one for each constraint of the original LP. For clarity, let's label these as WH, MH and PM. And it will have two constraints, associated with the variables of primal $P1$ and $P2$. Applying the rules of the Table 2.2 the dual can be obtained easily:

$$\text{MIN } W = 2700WH + 850MH + 95PM$$

$$\text{P1)} \ 30WH + 5MH + PM \geq 20$$

$$\text{P2)} \ 20WH + 10HM + PM \geq 60$$

$$WH, \ HM, \ \geq 0, \ PM \leq 0$$

2.4.2 Properties of the primal-dual relationship

There are some relevant properties concerning primal and dual:

The dual of dual is the primal

This can be easily proved just transforming the dual into a MAX canonical form and finding its dual. This means that duality defines a one-to-one correspondence between linear programs.

Optimum of primal and dual

An interesting property of duality is that if a linear program has a bounded optimum, its primal has also a bounded optimum and both have the same value:

$$z^* = w^* \tag{2.1}$$

Dual variables as shadow prices

Furthermore, the values of the dual variables in the optimum represent the *shadow price* of the constraints of the primal. This means that u_i^* is equal to:

$$u_i^* = \frac{\Delta z^*}{\Delta b_i} \tag{2.2}$$

That is, the value of the dual in the optimum u_i^* is equal to the change of the value of the optimum of the objective function divided by the change of the value of the right side term of its corresponding constraint

i in the primal. Sometimes it is said that u_i^* is the *shadow price* of constraint i.

As the dual of the dual is the primal, we can also write:

$$x_j^* = \frac{\Delta w^*}{\Delta c_j} = \frac{\Delta z^*}{\Delta c_j} \tag{2.3}$$

That is, the change of the value of the objective function in the optimum relative to the change of the cost coefficient c_j is equal to x_j^*.

2.5   Integer and mixed integer linear programming

The formulation of linear programming of section 2.2 states implicitly that variables x_j are real. But for some models it may be required that all decisions variables are integer: then we have *integer linear programming* (ILP). In other occasions, only a subset of the decision variables is required to be integer: that is an instance of *mixed integer linear programming* (MILP). Sometimes we will refer to MILP only when speaking of ILP and MILP, since the later category is more generic.

A special case of integer variables are *binary variables*, integer variables that can take only 0 and 1 values. Using binary variables widens considerably the possibilities of linear programming model building. Through binary variables can be modeled decision-making processes, and *logical constraints* can be introduced.

A first step to solve a MILP or ILP is solving its relaxed form. The *relaxed MILP* is a LP with the same objective function and constraints where all decision variables are real or continuous. If the integer variables of a MILP are integer in the optimum of the relaxed MILP, then the solution of the MILP is the same as the relaxed LP. There are some LP where the optimal solution is integer. A particular interesting subset satisfies the following properties:

- All righthand side values b_i are integer

- The constraint coefficients matrix $\mathbf{A}$ is totally unimodular

A matrix $\mathbf{A}$ is *totally unimodular* when any square submatrix of $\mathbf{A}$ (sometimes called minor) has determinant -1, 0 or $+1$. Some generic PL problems have this property, like the transportation problem (see subsection 2.2.3) or the assignment problem (see section 3.8).

For PL not satisfying this property, more generic strategies have to be developed. The *branch and bound* procedure was introduced by Ailsa H Lang and Alison G Doig as soon as 1960 [5]. Later the *cutting plane* and the *branch and cut* strategies were introduced. All these strategies start from the relaxation of the MILP, which provides a lower or upper bound (for the MIN and MAX problems, respectively) of the optimal value of the objective function. Later on, several linear programs are defined by adding constraints to the initial relaxed linear program, in order to find the solution of the MILP.

There are some relevant properties concerning MILP:

- The value of the objective function of the MILP will be poorer that the relaxed LP: smaller for MAX problems, bigger for MIN problems

- The resolution of the MILP can take more computational effort than the relaxed LP, as several LPs have to be solved.

- The results concerning duality and sensibility analysis obtained from the relaxed MILP are not applicable to MILP problems.

2.6 Solving linear programming in R

There are several solvers available for solving linear programming models. A list can be found in http://bit.ly/1zkJpVw. Some of these solvers can be embedded into larger programs to develop optimization problems. Some of them are written as C callable libraries, and are also implemented in R packages. The following packages can be of interest for R users:

- **lp_solve** is implemented through the **lpSolve** and **lpSolveAPI** packages

- GLPK is implemented through the **Rglpk** package

- SYMPHONY is implemented through **Rsymphony**

All solver are implemented as R functions, and parameters can be passed to these functions as R matrices and vectors. This also allows to embed these solvers into larger programs. Some of these packages have functions that can read LP and MILP programs from files, written in standards such as CPLEX, MPS or AMPL/MathProg. In all problems developed in chapter 3 there is a section dedicated to the code used to enter these models, and other section for the numerical results.

Most R packages solving LP implement solvers as functions, whose input variables are:

- A character variable indicating if we have a maximization or minimization problem

- Vectors with cost coefficients **c** and righthand side values **b**

- A matrix with the **A** coefficients

- A character vector with the constraint signs.

For ILP or MILP models, an additional vector indicating which variables are integer must be passed to the function. Alternatively, some logical variables indicate if all variables are integer or binary.

2.6.1 Solving two LPs with the lpSolve package

In small problems, like the one defined in subsection 2.2.2, the definition of parameters is easy, if we know something about the R notation. The following code solves that LP with two variables.

```
library(lpSolve)

#defining parameters

obj.fun <- c(20, 60)
constr <- matrix(c(30, 20, 5, 10, 1, 1), ncol = 2, byrow=
    TRUE)
constr.dir <- c("<=", "<=", ">=")
rhs <- c(2700, 850, 95)

#solving model

prod.sol <- lp("max", obj.fun, constr, constr.dir, rhs,
    compute.sens = TRUE)
```

```
#accessing to R output

prod.sol$obj.val #objective function value
prod.sol$solution #decision variables values
prod.sol$duals #includes duals of constraints and reduced
    costs of variables


#sensibility analysis results


prod.sol$duals.from
prod.sol$duals.to
prod.sol$sens.coef.from
prod.sol$sens.coef.to
```

For larger problems, there can be more efficient ways of passing model parameters than listing all variables. This is the case of the LP defined in subsection 2.2.3, a small instance of the more generic transportation problem. The following code defines the matrix **A** for any number of origins m and destinations n of a transportation problem.

```
library(lpSolve)

#defining parameters
#origins run i in 1:m
#destinations run j in 1:n
obj.fun <- c(8, 6, 3, 2, 4, 9)

m <- 2
n <- 3

constr <- matrix(0, n+m, n*m)

for(i in 1:m){
    for(j in 1:n){
```

```
15        constr[i, n*(i-1) + j] <- 1
          constr[m+j, n*(i-1) + j] <- 1
     }
}

20 constr.dir <- c(rep("<=", m), rep(">=", n))

   rhs <- c(70, 40, 40, 35, 25)

   #solving LP model
25 prod.trans <- lp("min", obj.fun, constr, constr.dir, rhs,
       compute.sens = TRUE)

   #LP solution
   prod.trans$obj.val
   sol <- matrix(prod.trans$solution, m, n, byrow=TRUE)
30 prod.trans$duals

   #sensitivity analysis of LP
   prod.trans$duals.from
   prod.trans$duals.to
35 prod.trans$sens.coef.from
   prod.trans$sens.coef.to
```

2.6.2   Syntax to parse LP models

When used outside R, PL solvers load the problems using several PL syntax. Among the most used syntaxs are CPLEX, MPS or MathProg. The following code picks a model written in CPLEX format, and uses the Rglpk package to solve it. It returns the solution in the original Rglpk format, and in data frame and LaTeX formats. It has been used to solve several LPs of the next chapter.

```
SolverLP <- function(model, method="CPLEX_LP", decimal=0)
    {
library(Rglpk)

model1.lp <- Rglpk_read_file(model, type = method,
    verbose=F)

model1.lp.sol <- Rglpk_solve_LP(model1.lp$objective,
    model1.lp$constraints[[1]], model1.lp$constraints
    [[2]], model1.lp$constraints[[3]], model1.lp$bounds,
    model1.lp$types, model1.lp$maximum)

library(xtable)

model1.lp.sol.df <- as.data.frame(model1.lp.sol$solution)
model1.lp.sol.df <- rbind(model1.lp.sol.df, c(model1.lp.
    sol$optimum))

rownames(model1.lp.sol.df) <- c(attr(model1.lp, "
    objective_vars_names"),"obj")
colnames(model1.lp.sol.df) <- "Solution"
table.sol <- xtable(model1.lp.sol.df, digits=decimal)
results <- list(sol=model1.lp.sol, df=model1.lp.sol.df,
    latex=table.sol)

return(results)
}
```

CHAPTER 3

Modeling linear programming

3.1 A production plan with fixed costs

A manufacturing manager is in charge of minimizing the total costs (raw materials, labor and storage costs) of the following four months. In Table 3.1 can be found the cost of raw materials of one unit of final product, the demand of final product and the working hours available for each month. Labor costs are of 12 € per hour, and only worked hours are payed. Each unit of final product needs 30 minutes of labor. Storage costs are equal to 2 € for each unit stored at the end of the month. Any unit produced at a given month can be used to cover the demand of the same month, or be stored to cover the demand of months to come. At the beginning of month 1 there is no stock, and there are no minimum stock requirements for any month.

Month	1	2	3	4
Unit cost (€)	6	8	10	12
Demand (units)	100	200	150	400
Working hours available	200	200	150	150

Table 3.1: Information for the production plan

1. Define the decision variables (provide a brief definition of each set of defined variables), objective function and constraints of a linear programming model that minimizes total production costs.

2. Modify the model of the previous section if a **fixed cost** of 1,000 € has to be taken into account for each month that there is production. This cost is assumed only if there is production different from zero in that month.

Models

1. Define the decision variables (provide a brief definition of each set of defined variables), objective function and constraints of a linear programming model that minimizes total production costs.

The variables used in to define the model are defined for $i = 1, \ldots, 4$:

- Variables q_i representing the quantity produced in month i

- Variables s_i representing the stock at the end of month i

The constraints d_i ensure that the demand is covered and constraints u_i should be added to make q_i no larger that its required upper bound.

$$\text{MAX } z = \sum_{i=1}^{4} (12q_i + 2s_i)$$

$$\text{d1) } q_1 - s_1 = 100$$

$$\text{d2) } s_1 + q_2 - s_2 = 200$$

$$\text{d3) } s_2 + q_3 - s_3 = 150$$

$$\text{d4) } s_3 + q_4 - s_4 = 400$$

$$\text{u1) } q_1 \leq 400$$

$$\text{u2) } q_2 \leq 400$$

$$\text{u3) } q_3 \leq 300$$

$$\text{u4) } q_4 \leq 300$$

$$s_i \geq 0$$

2. Modify the model of the previous section if a **fixed cost** of 1,000 € has to be taken into account for each month that there is production. This cost is assumed only if there is production different from zero in that month.

For this version of the model, four binary variables b_i are added, which equal one if there is production in month i, and zero otherwise. A set of constraints of the kind $q_i \leq Mb_i$ have been defined, although the constraints of upper bound can be also used, for instance making $q_1 \leq 400b_1$:

$$\text{MAX } z = \sum_{i=1}^{4} (12q_i + 2s_i + 1000b_i)$$

$$\text{d1) } q_1 - s_1 = 100$$

$$\text{d2) } s_1 + q_2 - s_2 = 200$$

$$\text{d3) } s_2 + q_3 - s_3 = 150$$

$$\text{d4) } s_3 + q_4 - s_4 = 400$$

$$\text{u1) } q_1 \leq 400b_1$$

$$\text{u2) } q_2 \leq 400b_2$$

$$\text{u3) } q_3 \leq 300b_3$$

$$\text{u4) } q_4 \leq 300b_4$$

$$s_i \geq 0, \; b_i \text{ binary}$$

Code

The CPLEX format of both models are:

```
Minimize
   cost:  12q1 + 14q2 + 16q3 + 18q4 + 2s1 + 2s2 + 2s3 + 2s4
Subject To
   d1:  q1 - s1 = 100
   d2:  s1 + q2 - s2 = 200
   d3:  s2 + q3 - s3 = 150
   d4:  s3 + q4 - s4 = 400
Bounds
   0 <= q1 <= 400
   0 <= q2 <= 400
   0 <= q3 <= 300
   0 <= q4 <= 300
End
```

```
Minimize
  cost: 12q1 + 14q2 + 16q3 + 18q4 + 2s1 + 2s2 + 2s3 + 2s4 +
        1000b1 + 1000b2 + 1000b3 + 1000b4
Subject To
  d1: q1 - s1 = 100
  d2: s1 + q2 - s2 = 200
  d3: s2 + q3 - s3 = 150
  d4: s3 + q4 - s4 = 400
  l1: q1 - 400b1 <= 0
  l2: q2 - 400b2 <= 0
  l3: q3 - 300b3 <= 0
  l4: q4 - 300b4 <= 0
Binary
  b1
  b2
  b3
  b4
End
```

Numerical solution

The solution of the proposed models can be found in Table 3.2 and Table 3.3.

	Month 1	Month 2	Month 3	Month 4
q_i	100	200	250	300
s_i	0	0	100	0

Table 3.2: Solution model 1 ($z = 13,600$)

	Month 1	Month 2	Month 3	Month 4
q_i	400	0	150	300
s_i	300	100	100	0
b_i	1	0	1	1

Table 3.3: Solution model 2 ($z = 16,600$ €)

3.2 A purchase plan with decreasing unit costs

A manufacturing manager is in charge of minimizing the purchasing costs (raw materials plus storage costs) of the following four months. In Table 3.4 can be found the cost of one unit of raw material and the demand of raw material for each month. Storage costs are equal to 2 € for each unit stored at the end of the month. Any unit of raw material purchased at given month can be used to cover the demand of the same month, or be stored to cover the demand of months to come. At the beginning of month 1 there is no stock, and there are no minimum stock requirements for any month.

Month	1	2	3	4
Unit cost (€)	12	14	16	18
Demand (units)	150	200	250	150

Table 3.4: Information for the purchasing plan

For the next four months, the supplier of raw materials has made an special offer: all units purchased above 200 in any given month will have a discounts of 2 €. For instance, if a purchase of 350 units is ordered in month 1, the first 200 units will be sold for 12 € each, and the following 150 will be sold for 10 € each.

1. Define the decision variables (provide a brief definition of each set of defined variables), objective function and constraints of a linear programming model that minimizes total purchasing costs.

Models

The challenge of this model is to make the linear program pick the first 200 expensive units of each month, before picking the cheap units. A possible way of doing so is to define the following variables for $i = 1, \ldots, 4$:

- Variables q_i representing the quantity purchased in month i equal or below 200

- Variables r_i representing the quantity purchased in month i above 200

- Variables s_i representing the stock at the end of month i

- Variables b_i binary which are equal to 1 if more than 200 units are purchased on month i

Note that the total purchase in a given month is equal to $q_i + r_i$. So picking the monthly demand from Table 3.4 we can define the constraints (where d_i is the demand listed on Table 3.4):

$$s_{i-1} + q_i + r_i - s_i = d_i$$

To be sure that we pick the expensive units before the cheap, we need to define the following constraints for each month:

$$q_i \leq 200$$
$$q_i \geq 200 b_i$$
$$r_i \leq M b_i$$

So if $b_i = 0$, we have that $q_i \leq 200$ and $r_i = 0$, since the second constraint is inactive. But when $b_i = 1$, we have that $q_i \leq 200$ and

$q_i \geq 200$ at the same time, thus $q_i = 200$, while there is no upper bound for r_i, if M is large enough.

Therefore, if c_i are the unit costs of purchasing on month i the model is:

$$\text{MIN } z = \sum_{i=1}^{4} \left(c_i q_i + \left(c_i - 2 \right) r_i + 2 s_i \right)$$

$$s_{i-1} + q_i + r_i - s_i = d_i \qquad\qquad i = 1, \ldots, 4$$

$$q_i \leq 200$$

$$q_i \geq 200 b_i$$

$$r_i \geq M b_i$$

$$q_i, \ r_i \geq 0, \ b_i$$

Code

A possible implementation of this model in CPLEX can be:

```
Minimize
  cost: 12q1 + 14q2 + 16q3 + 18q4 + 10r1 + 12r2 + 14r3 + 16
      r4 + 2s1 + 2s2 + 2s3 + 2s4
Subject To
  d1: q1 +r1 - s1 = 150
  d2: s1 + q2 + r2 - s2 = 200
  d3: s2 + q3 + r3 - s3 = 250
  d4: s3 + q4 + r4 - s4 = 150
  l1: q1 - 200b1 >= 0
  l2: q2 - 200b2 >= 0
  l3: q3 - 200b3 >= 0
  l4: q4 - 200b4 >= 0
  m1: r1 - 10000b1 <= 0
  m2: r2 - 10000b2 <= 0
  m3: r3 - 10000b3 <= 0
  m4: r4 - 10000b4 <= 0
Bounds
  0 <= q1 <= 200
  0 <= q2 <= 200
  0 <= q3 <= 200
  0 <= q4 <= 200
Binary
  b1
  b2
  b3
  b4
End
```

Numerical solution

In Table 3.5 is listed the solution of the model. The total costs of the production plan are of 10,200 €, and the best option is to purchase all units on month 1. The total amount to purchase on that month is $q_1 + r_1 = 200 + 550 = 750$.

	Month 1	Month 2	Month 3	Month 4
q_i	200	0	0	0
r_i	550	0	0	0
s_i	600	400	150	0
b_i	1	0	0	0

Table 3.5: Solution of problem 3.2. Total costs: 10,200 €

3.3 A production plan with extra capacity

You are in charge of planning the production of a chemical product for the next four months. The monthly demand and the purchasing unit costs of raw material are listed in Table 3.6. The capacity of the plant is of 1,300 tonnes (t.) per month. The demand of a month can be covered with the production of the same month, and also with production of past months. The storage costs are of 2 k€ per tonne stocked at the end of the month. The stock of finished product at the beginning of the first month is of 200 T, and it is expected to hold the same quantity at the end of the fourth month. There are no stocks of raw material, so all stocks are of finished product.

Month	1	2	3	4
Costs (k€/t)	3	8	6	7
Demand (t)	800	900	1,200	1,800

Table 3.6: Demand and unit production costs for the next following months

1. Obtain the linear programming model that allows to obtain the production plan which minimizes the sum of production and storage costs.

2. What is the meaning of the dual variables of the constraints defined in the model?

As the demand is proven to be irregular, the plant management is considering the possibility of adding extra capacity to the plant, introducing a new shift. This new shift would increase plant capacity in 400 T per month, but also would include an extra fixed cost of 500 k€. For legal reasons, it is not possible to add extra capacity in a month if it has been added in the previous month.

3. Modify the model obtained previously to include the possibility of including extra shifts, and assess the practicality of adding extra shifts.

Models

1. Obtain the linear programming model that allows to obtain the production plan which minimizes the sum of production and storage costs.

The variables to use in the model are:

- Variables q_i real: number of tonnes to produce on month i

- Variables s_i real: number of tonnes in stock at the end of month i

Then the model is as follows:

$$\text{MIN } z = 3q_1 + 8q_2 + 6q_3 + 7q_4 + 2\left(s_1 + s_2 + s_3 + s_4\right)$$

$$\text{s.a. D1) } 200 + q_1 = 800 + s_1$$

$$\text{D2) } s_1 + q_2 = 900 + s_2$$

$$\text{D3) } s_2 + q_3 = 1200 + s_3$$

$$\text{D4) } s_3 + q_4 = 1800 + s_4$$

$$\text{S4) } s_4 = 200$$

$$\text{C1) } q_1 \leq 1300$$

$$\text{C2) } q_2 \leq 1300$$

$$\text{C3) } q_3 \leq 1300$$

$$\text{C4) } q_4 \leq 1300$$

$$q_i \geq 0, \ s_i \geq 0$$

2. What is the meaning of the dual variables of the constraints defined in the model?

The dual variables are the *shadow price* of the constraint, that is, the variation of the objective function z caused by variations of the right-hand side term b_i of constraint i that do not change the optimal base. As the objective function represents the total costs, the meaning of the dual variables of the constraints is:

- For *constraints D1 to D4, and S4* this variable represents the *increase* of total costs as the demand of the considered month increases. Although formally the dual variable of these constraints is of unrestricted sign, it will be always *nonnegative*.

- For *constraints C1 to C4* this variable represents the *decrease* of total costs as the capacity of a given month increases. The dual variables of these constraints will be *nonpositive*.

3. Modify the model obtained previously to include the possibility of including extra shifts, and assess the practicality of adding extra shifts.

To consider the possibility of adding extra capacity to the model, a new set of binary variables has to be defined:

- Variables b_i that are equal to 1 if extra capacity is added in month i, and 0 otherwise.

These variables can allow us to include the constraints about the impossibility of contracting extra capacity in two consecutive months. Let's consider months 1 and 2, to begin with. The possible values that binary variables can have are listed in Table 3.7.

b_1	b_2	
0	0	True
0	1	True
1	0	True
1	1	False

Table 3.7: Possible values of variables b_1 and b_2

The only possibility we need to exclude from Table 3.7 is that $b_1 = 1$ and $b_2 = 1$ simultaneously. We can achieve this easily adding the constraint:

$$b_1 + b_2 \leq 1$$

We will proceed similarly for months 2 and 3, and months 3 and 4. So the final model is:

$$\text{MIN } z = 3q_1 + 8q_2 + 6q_3 + 7q_4 + 2\left(s_1 + s_2 + s_3 + s_4\right) +$$
$$+ 500\left(b_1 + b_2 + b_3 + b_4\right)$$

s.a. D1) $200 + q_1 = 800 + s_1$

D2) $s_1 + q_2 = 900 + s_2$

D3) $s_2 + q_3 = 1200 + s_3$

D4) $s_3 + q_4 = 1800 + s_4$

S4) $s_4 = 200$

C1) $q_1 \leq 1300 + 400b_1$

C2) $q_2 \leq 1300 + 400b_2$

C3) $q_3 \leq 1300 + 400b_3$

C4) $q_4 \leq 1300 + 400b_4$

B1) $b_1 + b_2 \leq 1$

B1) $b_2 + b_3 \leq 1$

B1) $b_3 + b_4 \leq 1$

$q_i \geq 0, \ s_i \geq 0, b_i \in \{0, 1\}$

Code

Below are listed the implementations of the defined models in CPLEX standard.

```
Minimize
   cost: 3q1 + 8q2 + 6q3 + 7q4 + 2s1 + 2s2 + 2s3 + 2s4
Subject To
   d0: s0 = 200
   d1: q1 + s0 - s1 = 800
   d2: s1 + q2 - s2 = 900
   d3: s2 + q3 - s3 = 1200
   d4: s3 + q4 - s4 = 1800
   d5: s4 = 200
Bounds
   0 <= q1 <= 1300
   0 <= q2 <= 1300
   0 <= q3 <= 1300
   0 <= q4 <= 1300
End
```

```
Minimize
  cost: 3q1 + 8q2 + 6q3 + 7q4 + 2s1 + 2s2 + 2s3 + 2s4 + 500
       b1 + 500b2 + 500b3 + 500b4
Subject To
  d0: s0 = 200
  d1: q1 + s0 - s1 = 800
  d2: s1 + q2 - s2 = 900
  d3: s2 + q3 - s3 = 1200
  d4: s3 + q4 - s4 = 1800
  d5: s4 = 200
  e1: q1 - 400b1 <= 1300
  e2: q2 - 400b2 <= 1300
  e3: q3 - 400b3 <= 1300
  e4: q4 - 400b4 <= 1300
  c1: b1 + b2 <= 1
  c2: b2 + b3 <= 1
  c3: b3 + b4 <= 1
Binary
  b1
  b2
  b3
  b4
End
```

Numerical solutions

The solution of the first model is listed in Table 3.8:

Month	1	2	3	4
q_i	1,300	800	1,300	1,300
s_i	700	600	700	200

Table 3.8: Optimal production plan (version 1). Total costs: 31,600 k€

The solution of the second model is listed in Table 3.9. If we compare the value of the objective function of both models, we can see that a saving is obtained including the possibility of adding extra capacity on months 1 and 4.

Month	1	2	3	4
q_i	1,700	0	1,300	1,700
s_i	1,100	200	300	200
b_i	1	0	0	1

Table 3.9: Optimal production plan (version 2). Total costs: 29,400 k€

3.4 Transportation by trucks

In Table 3.10 can be found the quarterly demand (in tonnes) and the acquisition costs per tonne (in k€ per tonne) for each quarter of raw materials for a chemical plant. All purchases in a given quarter can be used to cover the demand of the present quarter, or the demand of quarters in the future. The costs of stocking are of 8 k€ per tonne stored at the end of each quarter. The stocks at the beginning of first quarter are of 100 tonnes, and it is needed the same amount of stock at the end of the fourth quarter.

Quarter	T1	T2	T3	T4
Demand	1,000	1,200	1,500	1,800
Unit costs	20	25	30	40

Table 3.10: Demand of raw material (t) and unit costs (k€/t) for each quarter

In addition to the purchase and storage costs, the transportation costs have to be considered. All the purchased quantity of raw materials has to be transported, using any combination of the two available truck models:

- Small trucks: cost of 700 k€, and capacity of 500 tonnes.

- Large trucks: cost of 1,400 k€, and capacity of 1,200 tonnes.

We need to define a linear programming model that allows the minimization of the total costs: acquisition, storage and transport, obtaining the amount raw materials to purchase, and the amount of trucks of both kinds to be contracted each quarter.

Models

The variables to define are:

- q_i continuous: tonnes of raw material to purchase in quarter i

- s_i continuous: tonnes in stock at the end of quarter i, and s_0 as the initial stock

- t_i integer: small trucks to contract in quarter i

- u_i integer: large trucks to contract in quarter i

Once defined the variables, two sets of constraints have to be defined:

- Constraints assuring that the purchase plan meets the demand commitments. These are of the form $s_{i-1} + q_i - s_i = d_i$, being d_i the demand of the quarter.

- Constraints assuring that a sufficient number of each kind of trucks is contracted: $q_i \leq 500t_i + 1200u_i$

The resulting model is:

$$\text{MIN } z = \sum_{i=1}^{4} (c_i q_i + 8s_i + 700t_i + 1400u_i)$$

$$s_{i-1} + q_i - s_i = d_i \qquad\qquad i = 1, \ldots, 4$$

$$q_i - 500t_i - 1200u_i \leq 0 \qquad\qquad i = 1, \ldots, 4$$

$$s_0 = s_4 = 100$$

$$s_i, q_i \geq 0$$

$$t_i, u_i \geq 0, \text{integer}$$

where c_i and d_i are the unit costs and demand for each quarter reported in Table 3.10.

Code

```
Minimize
   cost: 20q1 + 25q2 + 30q3 + 40q4 + 8s1 + 8s2 + 8s3 + 8s4 +
         700t1 + 700t2 + 700t3 + 700t4 + 1400u1 + 1400u2 +
         1400u3 + 1400u4
Subject To
      sini:  s0 = 100
      dem1:  s0 + q1 - s1 = 1000
      dem2:  s1 + q2 - s2 = 1200
      dem3:  s2 + q3 - s3 = 1500
      dem4:  s3 + q4 - s4 = 1800
      sfin:  s4 = 100


      cap1:  q1 - 500t1 - 1200u1 <= 0
      cap2:  q2 - 500t2 - 1200u2 <= 0
      cap3:  q3 - 500t3 - 1200u3 <= 0
      cap4:  q4 - 500t4 - 1200u4 <= 0

Integer
      t1
      t2
      t3
      t4
      u1
      u2
      u3
      u4

End
```

Numerical solution

This is the numerical solution of the proposed model:

	Q1	Q2	Q3	Q4
q_i	900	1,200	3,400	0
s_i	0	0	1,900	100
t_i	2	0	0	0
u_i	0	1	3	0

Table 3.11: Solution of the linear program ($z = 173,000$ k€)

3.5 Production of two models of chairs

A company produces two models of chairs: 4P and 3P. The model 4P needs 4 legs, 1 seat and 1 back. On the other hand, the model 3P needs 3 legs and 1 seat. The company has a initial stock of 200 legs, 500 seats and 100 backs. If the company needs more legs, seats and backs, it can buy standard wood blocks, whose cost is 80 € per block. The company can produce 10 seats, 20 legs and 2 backs from a standard wood block. The cost of producing the model 4P is 30 €/chair, meanwhile the cost of the model 3P is 40 €/chair. Finally, the company informs that the minimum number of chairs to produce is 1,000 units per month.

1. Define a linear programming model, which minimizes the total cost (the production costs of the two chairs, plus the buying of new wood blocks).

Due to the economic crisis, the company has considered the possibility to just produce a single chair model between 3P and 4P.

2. Define the new linear programming model for producing only a single chair model, which minimizes the total cost.

Finally, the new CEO (Chief Executive Officer) of the company has decided that the factory needs to produce of the model 4P a minimum of 4 times the quantity of the model 3P.

3. Define the new linear programming model, which minimizes the total cost when producing 4P four times the quantity of 3P.

Models

1. Define a linear programming model, which minimizes the total cost (the production costs of the two chairs, plus the buying of new wood blocks).

The definition of variables is straigthforward:

- $X4P$: Number of chairs to produce of the model 4P

- $X3P$: Number of chairs to produce of the model 3P

- $XWOOD$: Number of standard wood block to buy

The required LP model is:

$$[MIN]cost = 80XWOOD + 30X4P + 20X3P$$

$$SEATS)X4P + X3P \leq 500 + 10XWOOD$$

$$LEGS)4X4P + 3X3P \leq 200 + 20XWOOD$$

$$BACKS)X4P \leq 100 + 2WOOD$$

$$DEM)X4P + X3P \geq 1000$$

where $X4P, X3P, XWOOD$ are integer and non-negative variables. Note that for each block of wood, 10 units of seats *and* 20 units of legs *and* 2 units of backs are produced.

2. Define the new linear programming model for producing only a single chair model, which minimizes the total cost.

This model includes the same variables as the model above, plus a new binary variable representing the decision of choosing between the 3P and the 4P model:

- BX: '1' means the factory has decided to produce the model 3P. '0' means the factory has decided to produce the model 4P. Binary.

$$[MIN]cost = 80XWOOD + 30X4P + 20X3P$$
$$SEATS)X4P + X3P \leq 500 + 10XWOOD$$
$$LEGS)4X4P + 3X3P \leq 200 + 20XWOOD$$
$$BACKS)X4P \leq 100 + 2WOOD$$
$$DEM)X4P + X3P \geq 1000$$
$$SEL\_3P)X3P \leq M \cdot BX$$
$$SEL\_4P)X4P \leq M \cdot (1 - BX)$$

where $X4P, X3P, XWOOD$ are integer and non-negative variables, BX is binary variable, and M is a large value, so the constraint SEL_3P is non active when $BX = 1$, and SEL_4P is non active when $BX = 0$.

3. Define the new linear programming model, which minimizes the total cost when producing 4P four times the quantity of 3P.

In this case, we don't need to add any new variable, but a constraint representing the restriction regarding the proportion between produced units of each chair.

$$[MIN]cost = 80XWOOD + 30X4P + 20X3P$$

$$SEATS)X4P + X3P \le 500 + 10XWOOD$$

$$LEGS)4X4P + 3X3P \le 200 + 20XWOOD$$

$$BACKS)X4P \le 100 + 2WOOD$$

$$DEM)X4P + X3P \ge 1000$$

$$TIMES)4X3P \le X4P$$

where $X4P, X3P, XWOOD$ are integer and non-negative variables.

Code

Below can be found the three models implemented in CPLEX standard:

```
Minimize
    cost: 80xwood + 30x4p + 40x3p
Subject To
    seats: x4p + x3p - 10xwood <= 500
    legs: 4x4p + 3x3p - 20xwood <= 200
    backs: x4p - 2xwood <= 100
    dem: x4p + x3p >= 1000
Integer
    x3p
    x4p
    xwood
End
```

In this second model, the BX variable has been labeled *decision*

```
Minimize
    cost: 80xwood + 30x4p + 40x3p
Subject To
    seats: x4p + x3p - 10xwood <= 500
    legs: 4x4p + 3x3p - 20xwood <= 200
    backs: x4p - 2xwood <= 100
    dem: x4p + x3p >= 1000
    dec3: x3p - 1000decision <= 0
    dec4: x4p + 1000decision <= 1000
Integer
    x3p
    x4p
    xwood
Binary
    decision
End
```

```
Minimize
     cost: 80xwood + 30x4p + 40x3p
Subject To
     seats: x4p + x3p - 10xwood <= 500
     legs: 4x4p + 3x3p - 20xwood <= 200
     backs: x4p - 2xwood <= 100
     dem: x4p + x3p >= 1000
     times: x4p - 4x3p >= 0
Integer
     x3p
     x4p
     xwood
End
```

Numerical solutions

In Table 3.12 can be found the numerical solution of the three models. Note that the model with less cost is the first one, since it is the one with a larger feasible region.

	Model 1	Model 2	Model 3
xwood	161	140	350
x4p	420	0	800
x3p	580	1,000	200
decision	—	1	—
obj(€)	48,680	51,200	60,000

Table 3.12: Solutions of the three proposed models

3.6 Hiring and firing

In Table 3.13 are listed the needs of pilots able to flight an A320 for the following six months. The cost of a pilot's salary is 8 k€ per month. At the beginning of Month 1 the airline has a staff of 20 pilots, but this staff can be adjusted each month.

Month	1	2	3	4	5	6
Needed pilots	30	60	55	40	45	50

Table 3.13: Needs of pilots for the following six months

Pilots can be hired and fired at the beginning of each month. Newly hired pilots can start working at the same month, and fired pilots stop working the same day they are fired. The cost of firing a pilot is 10 k€, and the hiring cost is of 5 k€ per pilot. If it is convenient, the airline can have a staff of pilots larger than the actual needs.

1. Define a linear programming model to obtain the pilots to hire and fire each month to minimize the total cost of pilot staff (costs of salary plus hiring and firing costs).

2. Modify the linear model to include the constraint that the airline cannot fire pilots if it has hired pilots the previous month.

Models

1. Define a linear programming model to obtain the pilots to hire and fire each month to minimize the total cost of pilot staff (costs of salary plus hiring and firing costs).

To model this situation, we'll have to define the following variables:

- Variables h_i: pilots hired at the beginning of month i

- Variables f_i: pilots fired at the beginning of month i

- Variables s_i: staff of pilots during month i

The model should have the following groups of constraints:

- Constraints assuring that the staff of pilots at the beginning of month i is equal to $s_i = h_i - f_i + s_{i-1}$. In this case, we have that $s_0 = 20$.

- Constraints assuring that variables s_i are bigger of equal to the values of staff required d_i listed in Table 3.13.

Then, the linear program to solve is:

$$[MIN]z = 5\sum_{i=1}^{6} h_i + 10\sum_{i=1}^{6} f_i + 8\sum_{i=1}^{6} s_i$$

$$s_i = h_i - f_i + s_{i-1} \qquad\qquad i = 1, \ldots, 6$$

$$s_i \geq d_i \qquad\qquad i = 1, \ldots, 6$$

$$h_i, f_i \geq 0 \qquad\qquad i = 1, \ldots, 6$$

The solution of this model can be found in Table 3.14.

2. Modify the linear model to include the constraint that the airline cannot fire pilots if it has hired pilots the previous month.

Looking at the solution of the previous problem in Table 3.14, it can be seen that this new constraint does not hold for months 2 and 3: in month 2 are hired 30 pilots, and in month 3 are fired 5 pilots. Then a new model has to be defined to account for this new restriction. To do so, we have to add a new binari variable:

- Variabe b_i: equals one if pilots are hired in month i, and zero otherwise

Then, two new sets of constraints must be added: one set assuring that $b_i = 0 \Rightarrow f_i = 0$, and another set making that $b_i = 1 \Rightarrow f_{i+1} = 0$:

$$[\text{MIN}]z = 5\sum_{i=1}^{6} h_i + 10\sum_{i=1}^{6} f_i + 8\sum_{i=1}^{6} s_i$$

$$s_i = h_i - f_i + s_{i-1} \qquad\qquad i = 1,\ldots,6$$

$$s_i \geq d_i \qquad\qquad i = 1,\ldots,6$$

$$f_i \leq Mb_i \qquad\qquad i = 1,\ldots,5$$

$$h_{i+1} \leq M(1 - b_i) \qquad\qquad i = 1,\ldots,5$$

$$b_i \in \{0,1\} \qquad\qquad i = 1,\ldots,5$$

$$h_i, f_i \geq 0 \qquad\qquad i = 1,\ldots,6$$

The solution for this new model is listed in Table 3.15.

Code

```
Minimize
   cost: 5h1 + 5h2 + 5h3 + 5h4 + 5h5 + 5h6 + 10f1 + 10f2 +
      10f3 + 10f4 + 10f5 + 10f6 + 8s1 + 8s2 + 8s3 + 8s4 + 8
      s5 + 8s6
Subject To
      sini: s0 = 20
      sm1: s0 + h1 - f1 - s1 = 0
      sm2: s1 + h2 - f2 - s2 = 0
      sm3: s2 + h3 - f3 - s3 = 0
      sm4: s3 + h4 - f4 - s4 = 0
      sm5: s4 + h5 - f5 - s5 = 0
      sm6: s5 + h6 - f6 - s6 = 0
Bounds
   30 <= s1
   60 <= s2
   55 <= s3
   40 <= s4
   45 <= s5
   50 <= s6
End
```

```
Minimize
   cost: 5h1 + 5h2 + 5h3 + 5h4 + 5h5 + 5h6 + 10f1 + 10f2 +
      10f3 + 10f4 + 10f5 + 10f6 + 8s1 + 8s2 + 8s3 + 8s4 + 8
      s5 + 8s6
Subject To
      sini: s0 = 20
      sm1: s0 + h1 - f1 - s1 = 0
      sm2: s1 + h2 - f2 - s2 = 0
      sm3: s2 + h3 - f3 - s3 = 0
      sm4: s3 + h4 - f4 - s4 = 0
      sm5: s4 + h5 - f5 - s5 = 0
      sm6: s5 + h6 - f6 - s6 = 0
      hf01: f1 - 1000b1 <= 0
```

```
      hf02:  f2 - 1000b2 <= 0
      hf03:  f3 - 1000b3 <= 0
      hf04:  f4 - 1000b4 <= 0
      hf05:  f5 - 1000b5 <= 0
      hf06:  h2 + 1000b1 <= 1000
      hf07:  h3 + 1000b2 <= 1000
      hf08:  h4 + 1000b3 <= 1000
      hf09:  h5 + 1000b4 <= 1000
      hf10:  h6 + 1000b5 <= 1000
Bounds
   30 <= s1
   60 <= s2
   55 <= s3
   40 <= s4
   45 <= s5
   50 <= s6
Binary
b1
b2
b3
b4
b5
End
```

Numerical solutions

Below are listed the solutions of both linear programs. In the second case the values of binary variables has been omitted.

Month	1	2	3	4	5	6
Hired	10	30	0	0	0	5
Fired	0	0	5	10	0	0
Staff	30	60	55	45	45	50
Staff req.	30	60	55	40	45	50

Table 3.14: *Optimal solution for the first model of staff planning. Total costs: 2,655 k€*

Month	1	2	3	4	5	6
Hired	10	30	0	0	0	5
Fired	0	0	0	15	0	0
Staff	30	60	60	45	45	50
Staff req.	30	60	55	40	45	50

Table 3.15: *Optimal solution for the second model of staff planning. Total costs: 2,695 k€*

3.7 Planning of shifts through linear programming

A company has a emergency center which is working 24 hours a day. In Table 3.16 is detailed the minimal needs of employees for each of the six shifts of four hours in which the day is divided.

Shift	Employees
00:00 - 04:00	5
04:00 - 08:00	7
08:00 - 12:00	18
12:00 - 16:00	12
16:00 - 20:00	15
20:00 - 00:00	10

Table 3.16: Information for the production plan

Each of the employees of the emergency center works eight hours a day, covering two consecutive shifts of four hours. For instance, a given employee may start working at 20:00, and end working at 04:00.

You are asked to define a linear programming model which can define a planning of shifts that allows to cover the minimal needs for each shift with a minimum number of employees.

Model

To define the model, a set of eight variables has to be defined:

- Variable s_i (integer): number of employees that starts working in shift i

Then, the model to be defined for this situation is:

$$[\text{MIN}] z = s_1 + s_2 + s_3 + s_4 + s_5 + s_6$$

$$s_6 + s_1 \geq 5$$

$$s_1 + s_2 \geq 7$$

$$s_2 + s_3 \geq 18$$

$$s_3 + s_4 \geq 12$$

$$s_4 + s_5 \geq 15$$

$$s_5 + s_6 \geq 10$$

$$s_i \text{ integer}$$

Note that the constraints have been defined as greater o equal: the data in Table 3.16 is interpreted as the *minimal* number of employees required for each shift. If constraints were defined as inequalities, there should be only one solution, which can be not integer. This interpretation gives flexibility to the model in order to find the optimal solution. In Table 3.17 is listed the solution for this model: all minimal needs are covered with a staff of 38 employees.

Code

```
Minimize
  workforce: s1 + s2 + s3 + s4 + s5 + s6
Subject To
  t1: s6 + s1 >= 5
  t2: s1 + s2 >= 7
  t3: s2 + s3 >= 18
  t4: s3 + s4 >= 12
  t5: s4 + s5 >= 15
  t6: s5 + s6 >= 10
Integer
  s1
  s2
  s3
  s4
  s5
  s6
End
```

Numerical solution

	Solution
s1	5
s2	6
s3	12
s4	0
s5	15
s6	0
obj	38

Table 3.17: Optimal solution of the proposed model (number of employees)

3.8 Assignment maximizing minimal quality

In Table 3.18 can be found the quality with which five teachers (T1 to T5) teach five courses (C1 to C5). Each teacher teaches one course and each course is taught by one teacher.

	C1	C2	C3	C4	C5
T1	34	87	26	47	76
T2	43	90	24	63	97
T3	60	65	64	83	54
T4	89	62	39	37	18
T5	27	15	69	93	96

Table 3.18: Quality of courses C when taught by teacher T

We intend to define two LP models to assign teachers to courses following two criteria of quality:

1. Maximizing the total quality of courses obtained from the assignment

2. Maximizing the minimal quality of courses obtained from the assignment

Models

Total quality maximization

This problem is an instance of the more generic assignment problem: to assign tasks (courses) to agents (teachers) to maximize total quality. This formulation is equivalent to maximizing average quality, since this average is equal to total quality divided by the number of tasks.

To solve this problem we need to define the variables:

- Variable x_{ij} binary, which equals one if task j is assigned to agent i and zero otherwise.

The cost coefficients of the objective function will be the elements c_{ij} of Table 3.18, and two groups of constraints are needed:

- Constraints assuring that each course j is taught by a one teacher only

- Constraints assuring that each teacher i is teaching only one course

The model is:

$$\text{MAX } z = \sum_{i=1}^{n} \sum_{j=1}^{n} c_{ij} x_{ij}$$

$$\sum_{i=1}^{n} x_{ij} = 1 \qquad\qquad j = 1, \ldots, n$$

$$\sum_{j=1}^{n} x_{ij} = 1 \qquad\qquad i = 1, \ldots, n$$

$$x_{ij} \text{ binary}$$

Maximization of minimal quality

This variant is a particular case of a maximin linear program formulation, that is, *maximizing the minimum value of a set of functions*. To implement this formulation, we need the same variables x_{ij} plus a variable q which will be a lower bound of course quality.

To make q a lower bound of course quality it must be stated that the quality of any course j will be greater or equal than q:

$$\sum_{i=1}^{n} c_{ij} x_{ij} \geq q \qquad\qquad j = 1, \dots, n$$

Then, to maximize minimal quality is equivalent to maximize variable q:

$$\text{MAX } z = q$$

$$\sum_{i=1}^{n} x_{ij} = 1 \qquad\qquad j = 1, \dots, n$$

$$\sum_{j=1}^{n} x_{ij} = 1 \qquad\qquad i = 1, \dots, n$$

$$\sum_{i=1}^{n} c_{ij} x_{ij} \geq q \qquad\qquad j = 1, \dots, n$$

$$x_{ij} \text{ binary}$$

The results of assigning teachers to courses following the criteria of maximizing total quality and maximizing minimal quality can be found in Table 3.19 and Table 3.20, respectively.

Code

In this case, it coud be preferable to develop a specific R function to solve an instance of any size. This function loads the elements of Rglpk_solve_LP to solve the first version of the problem (maximization of total course quality).

```r
Assignment01 <- function(c){
    n <- dim(c)[1]
    coef <- as.vector(t(c))
    rhs <- rep(1, 2*n)

5
    Amatrix <- matrix(0, 2*n, n*n)

    for(i in 1:n){
        for(j in 1:n){
10          Amatrix[i, n*(i-1)+j] <-1
        }
    }

    for(i in 1:n){
15      for(j in 1:n){
            Amatrix[n+i, n*(j-1)+i] <- 1
        }
    }

20  signs <- rep("==", 2*n)
    var_type <- rep("B", 2*n)
    library(Rglpk)

    solution <- Rglpk_solve_LP(obj=coef, mat=Amatrix, dir=
        signs, types=var_type, rhs=rhs, max=TRUE)
25  return(solution)
}
```

The implementation for the second model (maximization of minimal course quality) is the function:

```
Assignment02 <- function(c){
    n <- dim(c)[1]
    coef <- c(rep(0,n*n), 1)
    rhs <- c(rep(1, 2*n), rep(0, n))
    Amatrix <- matrix(0, 3*n, n*n + 1)

    for(i in 1:n){
        for(j in 1:n){
            Amatrix[i, n*(i-1)+j] <-1
        }
    }

    for(i in 1:n){
        for(j in 1:n){
            Amatrix[n+i, n*(j-1)+i] <- 1
        }
    }

    for(i in 1:n){
        for(j in 1:n){
            Amatrix[2*n+i, n*(j-1)+i] <- c[j, i]
        }
    }

    for(i in 1:n){
        Amatrix[2*n+i, n*n + 1] <- -1
    }

    signs <- c(rep("==", 2*n), rep(">=", n))

    var_type <- c(rep("B", n*n), "C")

    library(Rglpk)
```

```
35    solutionPL <- Rglpk_solve_LP(obj=coef, mat=Amatrix, dir
         =signs, types=var_type, rhs=rhs, max=TRUE)

      return(solutionPL)

}
```

To obtain the solutions of both models using the functions, we run the code below. Solutions can be found in Table 3.19 and Table 3.20, respectively.

```
#sample matrix has been generated at random

set.seed(1)
c <- matrix(sample(10:100, 25), 100, 100)

#running of the first model

solAss01 <- Assignment01(c)
m.01 <- matrix(solAss01$solution[1:25], 5, 5, byrow=TRUE)

#running of the second model

solAss02 <- Assignment02(c)

m.02 <- matrix(solAss$solution[1:25], 5, 5, byrow=TRUE)
```

	C1	C2	C3	C4	C5
T1	34	**87**	26	47	76
T2	43	90	24	63	**97**
T3	60	65	**64**	83	54
T4	**89**	62	39	37	18
T5	27	15	69	**93**	96

Table 3.19: Assignment to maximize total quality (in bold)

Both solutions give quite good assignments. The maximum total quality gives a solution with average quality of 86, while the maximum minimal quality criterion gives a solution with average quality equal to 81.4. But the while the first criterion has values of quality from 97 to 64,

	C1	C2	C3	C4	C5
T1	34	87	26	47	**76**
T2	43	**90**	24	63	97
T3	60	65	64	**83**	54
T4	**89**	62	39	37	18
T5	27	15	**69**	93	96

Table 3.20: Assignment to maximize minimal quality (in bold)

in the second criterion quality ranges from 90 to 69, assuring more homogeneity.

3.9   Production of biofuel

A company that produces aircraft biofuel is planning a new product called FC (Fuel-Corn). Table 3.21 shows the total quarterly demand in tonnes (t) for the coming years as communicated by their customers.

	Q1	Q2	Q3	Q4
FC demand (T)	1,200	1,100	1,300	1,000

Table 3.21: FC quarterly demand

In Table 3.22 can be found the costs per tonne of Fuel and Corn for every two month period in the years to come.

	B1	B2	B3	B4	B5	B6
Fuel (k€/t)	2	2.5	2	1	1.5	3
Corn (k€/t)	1.5	1	2	1	2	2.5

Table 3.22: Costs of Fuel and Corn in bimonthly periods

FC composition is obtained by mixing 35% of Fuel and 65% of Corn. The life of Fuel is of four consecutive months and the life of Corn, six (i.e., if we buy Fuel in early January, we cannot use it in early May). We just buy Fuel and Corn at the beginning of each two-month period and make the deliveries of FC at the beginning of each quarter. For simplicity, we assume that one can buy, mix and sell the same day.

In addition, the plant manager has told us that in any two-month period, we cannot buy more Fuel than triple of Corn.

In these conditions, you are required to:

- develop a model to determine the amount of Fuel and Corn to buy every two months to minimize the annual cost of production of FC.

- The representative of Corn has offered a discount through which, if in a two month period one buys 1,000 tons or more, they sell all the tonnes purchased with a discount of 25%.

- Furthermore, the representative of Fuel has imposed that if in a two-month period more than 400 tonnes of Fuel are bought, no Fuel can be purchased in the following two months.

NOTE: The models of the second and third situation are independent, and should be built starting from the first model.

Models

The point of this model is that raw materials are bought every two months, and final product dispatched every three months. After considering each case, in Table 3.23 we find the two month periods in which Fuel and Corn can be bought to cover the demand of each quarter, and in Table 3.24 the quarters where can be used Fuel and Corn bought on each period.

Quarter j	Fuel (set F_j)	Corn (set C_j)
1	6, 1	5, 6, 1
2	1, 2	6, 1, 2
3	3, 4	2, 3, 4
4	4, 5	3, 4, 5

Table 3.23: Purchase periods of raw materials for each quarter

Period i	Fuel (set F_i^{-1})	Corn (set C_i^{-1})
1	1, 2	1,2
2	2	2, 3
3	3	3, 4
4	3, 4	3, 4
5	4	4, 1
6	1	1, 2

Table 3.24: Purchase periods of raw materials for each quarter

Then, we define variables f_{ij} and c_{ij}, representing the amount of Fuel and Corn, respectively, to buy on period i to cover the demand of quarter j. The cost coefficients of the variables are the values q_i and r_i, respectively, of Table 3.22.

The model has three blocks of constraints:

- Two sets of constraints indicating the demand of raw materials for each quarter. The demand of Fuel and Corn is the 35% and 65% of quarterly demand indicated in Table 3.21, respectively.

- A set of constraints to control that we cannot buy more Fuel than triple of Corn for each period.

Then, the model can be formulated as:

$$\text{MIN } z = \sum_{i \in F_j} \sum_{j \in 1,\ldots,4} q_i f_{ij} + \sum_{i \in C_j} i \sum_{j \in 1,\ldots,4} r_i c_{ij}$$

$$\sum_{i \in F_j} f_{ij} \geq 0.35 d_i \qquad\qquad j = 1,\ldots,4$$

$$\sum_{i \in C_j} c_{ij} \geq 0.65 d_i \qquad\qquad j = 1,\ldots,4$$

$$\sum_{j \in F_i^{-1}} f_{ij} \leq 3 \sum_{j \in C_i^{-1}} c_{ij} \qquad\qquad i = 1,\ldots,6$$

$$f_{ij},\ c_{ij} \geq 0$$

The second model posits a varying purchase price, but with a different scheme as in problem 3.2. In this case, if we buy more than 1,000 tons of Corn, the 25% discount is applied to *all the tons of Corn purchased in that two-month period*. To model this situation, additional variables should be defined:

- Variables c_i: amount of Corn purchased on period i, if the total amount is *below* 1,000 tonnes.

- Variables e_i: amount of Corn purchased on period i, if the total amount is *above* 1,000 tonnes.

- Binary variable b_i, equal to one if more than 1,000 tonnes of Fuel are bought on period i and zero otherwise.

Then, the new model formulation is:

$$\text{MIN } z = \sum_{i \in F_j} \sum_{j \in 1,\ldots,4} q_i f_{ij} + \sum_{i \in 1,\ldots,6} r_i c_i + \sum_{i \in 1,\ldots,6} 0.75 r_i e_i$$

$$c_i + e_i = \sum_{j \in C_i^{-1}} c_{ij} \qquad\qquad\qquad i = 1,\ldots,6$$

$$\sum_{i \in F_j} f_{ij} \geq 0.35 d_i \qquad\qquad\qquad j = 1,\ldots,4$$

$$\sum_{i \in C_j} c_{ij} \geq 0.65 d_i \qquad\qquad\qquad j = 1,\ldots,4$$

$$\sum_{j \in F_i^{-1}} f_{ij} \leq 3 \sum_{j \in C_i^{-1}} c_{ij} \qquad\qquad i = 1,\ldots,6$$

$$d_i \geq 1000 b_i$$

$$c_i \leq M\left(1 - b_i\right)$$

$$d_i \leq M b_i$$

$$f_{ij},\ c_{ij},\ d_i,\ e_i \geq 0$$

$$b_i \text{ binary}$$

Finally, to introduce the constraints relative to fuel (if more than 400 tons are bought in the two-month period, no Fuel can be purchased in the following two-month period), we introduce binary variables k_i which equal one if more than 400 tons are bought in two month period i and zero otherwise. Variables f_j equaling the total amount of fuel purchased on i are also introduced.

$$\text{MIN } z = \sum_{i \in F_j} \sum_{j \in 1,\ldots,4} q_i f_{ij} + \sum_{i \in C_j} i \sum_{j \in 1,\ldots,4} r_i c_{ij}$$

$$f_i = \sum_{j \in F_i^{-1}} f_{ij} \qquad\qquad i = 1,\ldots,6$$

$$\sum_{i \in F_j} f_{ij} \geq 0.35 d_i \qquad\qquad j = 1,\ldots,4$$

$$\sum_{i \in C_j} c_{ij} \geq 0.65 d_i \qquad\qquad j = 1,\ldots,4$$

$$\sum_{j \in F_i^{-1}} f_{ij} \leq 3 \sum_{j \in C_i^{-1}} c_{ij} \qquad\qquad i = 1,\ldots,6$$

$$f_i \geq 400 k_i \qquad\qquad i = 1,\ldots,5$$

$$f_i \leq M(1 - k_{i-1}) \qquad\qquad i = 2,\ldots,6$$

$$f_i, \ f_{ij}, \ c_{ij} \geq 0 \qquad\qquad k_i \text{ binary}$$

Code

The first model in CPLEX format:

```
Minimize
    2F11 + 2F12 + 2.5F22 + 2F33 + F43 + F44 + 1.5F54 + 3F61
        + 1.5C11 + 1.5C12 + C22 + C23 + 2C33 + 2C34 + C43 +
        C44 + 2C54 + 2C51 + 2.5C61 + 2.5C62
Subject To
    fuel1: F61 + F11 >= 420
    fuel2: F12 + F22 >= 385
    fuel3: F33 + F43 >= 455
    fuel4: F44 + F54 >= 350

    corn1: C51 + C61 + C11 >= 780
    corn2: C62 + C12 + C22 >= 715
    corn3: C23 + C33 + C43 >= 845
    corn4: C34 + C44 + C54 >= 650

    prop1: F11 + F12 - 3C11 - 3C12 <= 0
    prop2: F22 - 3C22 - 3C23 <= 0
    prop3: F33 - 3C33 - 3C34 <= 0
    prop4: F43 + F44 - 3C43 - 3C44 <= 0
    prop5: F54 - 3C54 - 3C51 <= 0
    prop6: F61 - 3C61 - 3C62 <= 0
End
```

The second model in CPLEX format (e_i variables are defined as DI in this implementation):

```
Minimize
    2F11 + 2F12 + 2.5F22 + 2F33 + F43 + F44 + 1.5F54 + 3F61
        + 1.5C1 + C2 + 2C3 + C4 + 2C5 + 2.5C6 + 1.125D1 +
        0.75D2 + 1.5D3 + 0.75D4 + 1.5D5 + 1.875D6
```

```
Subject To
    vars1: C1 + D1 - C11 - C12 = 0
    vars2: C2 + D2 - C22 - C23 = 0
    vars3: C3 + D3 - C33 - C34 = 0
    vars4: C4 + D4 - C43 - C44 = 0
    vars5: C5 + D5 - C54 - C51 = 0
    vars6: C6 + D6 - C61 - C62 = 0

    fuel1: F61 + F11 >= 420
    fuel2: F12 + F22 >= 385
    fuel3: F33 + F43 >= 455
    fuel4: F44 + F54 >= 350

    corn1: C51 + C61 + C11 >= 780
    corn2: C62 + C12 + C22 >= 715
    corn3: C23 + C33 + C43 >= 845
    corn4: C34 + C44 + C54 >= 650

    prop1: F11 + F12 - 3C1 - 3D1 <= 0
    prop2: F22 - 3C2 - 3D2 <= 0
    prop3: F33 - 3C3 - 3D3 <= 0
    prop4: F43 + F44 - 3C4 - 3D4 <= 0
    prop5: F54 - 3C5 - 3D5 <= 0
    prop6: F61 - 3C6 - 3D6 <= 0

    des01: D1 - 1000B1 >= 0
    des02: D2 - 1000B2 >= 0
    des03: D3 - 1000B3 >= 0
    des04: D4 - 1000B4 >= 0
    des05: D5 - 1000B5 >= 0
    des06: D6 - 1000B6 >= 0

    Ces01: C1 + 1000B1 <= 1000
    Ces02: C2 + 1000B2 <= 1000
    Ces03: C3 + 1000B3 <= 1000
```

```
        Ces04:  C4 +  1000B4 <=  1000
        Ces05:  C5 +  1000B5 <=  1000
        Ces06:  C6 +  1000B6 <=  1000


        Des01:  D1 - 10000B1 <=  0
        Des02:  D2 - 10000B2 <=  0
        Des03:  D3 - 10000B3 <=  0
        Des04:  D4 - 10000B4 <=  0
        Des05:  D5 - 10000B5 <=  0
        Des06:  D6 - 10000B6 <=  0


Binary
        B1
        B2
        B3
        B4
        B5
        B6
End
```

The CPLEX formulation of the third model is as follows. Here the k_i variables have been labeled as BF.

```
Minimize
      2F11 +  2F12 +  2.5F22 +  2F33 +  F43 +  F44 +  1.5F54 +  3F61
          +  1.5C11 +  1.5C12 +  C22 +  C23 +  2C33 +  2C34 +  C43 +
          C44 +  2C54 +  2C51 +  2.5C61 +  2.5C62
Subject To

      Vars1:  F1 -  F11 -  F12 = 0
      Vars4:  F4 -  F43 -  F44 = 0

      fuel1:  F61 +  F11 >=  420
      fuel2:  F12 +  F22 >=  385
      fuel3:  F33 +  F43 >=  455
```

```
fuel4:  F44 + F54 >= 350


corn1:  C51 + C61 + C11 >= 780
corn2:  C62 + C12 + C22 >= 715
corn3:  C23 + C33 + C43 >= 845
corn4:  C34 + C44 + C54 >= 650


prop1:  F11 + F12 - 3C11 - 3C12 <= 0
prop2:  F22 - 3C22 - 3C23 <= 0
prop3:  F33 - 3C33 - 3C34 <= 0
prop4:  F43 + F44 - 3C43 - 3C44 <= 0
prop5:  F54 - 3C54 - 3C51 <= 0
prop6:  F61 - 3C61 - 3C62 <= 0


res01:  F1 - 400BF1 >= 0
res02:  F22 - 400BF2 >= 0
res03:  F33 - 400BF3 >= 0
res04:  F4 - 400BF4 >= 0
res05:  F54 - 400BF5 >= 0
res06:  F61 - 400BF6 >= 0


res13:  F22 - 90000BF1 <= 90000
res14:  F33 - 90000BF2 <= 90000
res15:  F4 - 90000BF3 <= 90000
res16:  F54 - 90000BF4 <= 90000
res17:  F61 - 90000BF5 <= 90000


Binary
     BF1
     BF2
     BF3
     BF4
     BF5
     BF6
End
```

Numerical results

In Table 3.25 are the numerical results for the first and third models. As in the first model Fuel is bought in periods 1 and 4, the third model has the same numerical results as the first. Table 3.26 shows the results of the second model. Note that in that second model more corn than needed is bought to benefit from the discount, but the total costs are lower than in the other two models.

	Solution
F11	420
F12	385
F43	455
F44	350
C11	780
C22	715
C23	845
C44	650
obj	5,795

Table 3.25: Results for the first and third models (variables equal to zero omitted). Variables in tonnes i obj. function in k€

	Solution
F11	420
F12	385
F43	455
F44	350
D1	1,000
D2	1,000
D4	1,000
C11	790
C12	210
C22	505
C23	495
C43	350
C44	650
B1	1
B2	1
B4	1
obj	5,040

Table 3.26: Results for the first and third models (variables equal to zero omitted)

3.10 A finantial optimization problem

A university student has a grant to work as intern in the Operations Research department of her university. She starts working in January, and receives 3,500 € at the end of each month. She has enough money to pay her bills this year, so she has decided to invest her money.

	0	4,000	20,000
1	3.04	3.56	3.82
2	3.24	3.76	4.04
3	3.44	3.96	4.26
4	3.64	4.16	4.48
5	3.84	4.36	4.70
6	4.04	4.56	4.92
7	4.25	4.75	5.15
8	4.45	4.95	5.37
9	4.65	5.15	5.59
10	4.85	5.35	5.81
11	5.05	5.55	6.03
12	5.25	5.75	6.25

Table 3.27: Yearly interests rates for every category

She has gone to the campus office bank, and she has been told that she can get interests from her money in the following conditions:

- She can contract a fixed deposit at the beginning of each month. The yearly interest rates are dependent upon the term of the deposit, and are listed in Table 3.27. Monthly interest rates can be obtained dividing the yearly rate by twelve.

- She can retrieve her money at the end of December, irrespective of the moment of time she has made the deposit. This means that, for instance, if she deposits money at the beginning of month four, she will have the money deposited during nine months and she will receive the interest corresponding to that deposit as listed in Table 3.27.

- As can be seen in Table 3.27, there are three interests depending of the amount of the deposit. Interests apply to the total amount of the diposit. For instance, if in month 2 are deposited 4,000 €, money will remain deposited for eleven months. So, she will receive a monthly interest of 5.05/12%. But if the amount of the deposit is between 4,000 and 20,000, then she will receive a monthly interest of 5.55/12% *on the total amount deposited*.

She has found that the conditions are quite adequate for her, so she has contracted the deposit. To start her savings plan, she counts with 3,500 € at the beginning of January. Given these conditions, you are requested to find through linear programming the amount to deposit each month to maximize the total interests earned at the end of month 12.

Models

It seems obvious that a set of variables representing the amount to deposit each month for every category must be defined:

- p_i: total amount deposited in the range of the first category (less than 4,000) in month i

- q_i: total amount deposited in the range of the second category (more than 4,000) and less than 20,000) in month i

- r_i: total amount deposited in the range of the third category (more than 20,000) in month i

where $i = 1, \ldots, 12$. Note that for each month, one one of the three variables can be different from zero.

The cost coefficients k_{ij} will be obtained from Table 3.27. If k_{ij} is the yearly interest offered for a deposit in the month i in category j we have that cost coefficients are equal to:

- $k_{i1}(12 + 1 - i)/12 = k_{i1}(13 - i)/12$ for variables p_i

- $k_{i2}(13 - i)/12$ for variables q_i

- $k_{i3}(13 - i)/12$ for variables r_i

so the objective function is:

$$\text{MAX } z = \sum_{i=1}^{12} \frac{13 - i}{12} \left(k_{i1}p_i + k_{13-i,2}q_i + k_{13-i,3}r_i \right)$$

In a given month, the total amount to be deposited will be equal to $p_i + q_i + r_i$. Money can come from the same month, or for previous months. So we need the variables:

- s_i money available, but not deposited at the end of month i

So the *continuity* constraints are:

$$s_{i-1} + 3500 = p_i + q_i + r_i + s_i$$

for $i = 1, \ldots, 12$, with $s_0 = 0$.

Finally, we must set the values of the variables to their corresponding category. For doing so, we must define two binary variables for each month:

- b_i: equals one if the money deposited in month i belongs to the second category, and zero otherwise

- c_i: equals one if the money deposited in month i belongs to the third category, and zero otherwise

If the money deposited belongs to the third category, both binary variables equal zero. As the interests of the third category are larger than the ones of the second for all money deposited, b_i and c_i never will equal one at the same time in the optimal solution.

Therefore, we must add the constraints:

$$q_i \leq 4000 b_i$$
$$r_i \geq 20000 c_i$$
$$p_i \leq M \left(1 - b_i\right) q_i \qquad\qquad \leq M \left(1 - c_i\right)$$

Code

Here is the code that implements the model described above. Data of interest rates is read from a .csv file, and loaded into the *fin* object.

```
library(Rglpk)

#defining the objective function

fin <- fin[,2:13]/12
for(i in 1:12) fin[,i] <- fin[,i]*i

int <- fin[,12:1]
f.obj <- c(t(int[1,]),t(int[2,]),t(int[3,]),rep(0,11))
l <- length(f.obj)
f.obj <- c(f.obj,rep(0,12*2))

#defining types of variables

types <- c(rep("C",l),rep("B",12*2))

#defining constraints
mat1 <- matrix(0,nrow=12,ncol=length(f.obj))

for(i in 1:12){
  mat1[i,c(i,12+i,24+i)] <- 1
  if(i>1) mat1[i,l+i-1-11] <- -1
  if(i<12) mat1[i,l+i-11] <- 1
}


f.rhs1 <- c(rep(3500,12))
f.dir1 <- rep("==",12)

#binary variables constraints (greater of equal)
mat2 <- matrix(0,nrow=12*2,ncol=length(f.obj))
```

```
for(i in 1:(12*2)){
  mat2[i,i+12] <- 1
  if(i<=12) mat2[i,i+1] <- -4000
  if(i>12) mat2[i,i+1] <- -20000
}

f.rhs2 <- c(rep(0,12*2))
f.dir2 <- rep(">=",12*2)

#binary variables constraints (lesser or equal)
M=1000000

mat3 <- matrix(0,nrow=12*2,ncol=length(f.obj))

for(i in 1:(12*2)){
  mat3[i,c(12+i)] <- 1
  mat3[i,1+i] <- -M
}

f.rhs3 <- c(rep(0,12*2))
f.dir3 <- rep("<=",12*2)

#binding all constraints
f.con <- rbind(mat1,mat2,mat3)
f.rhs <- c(f.rhs1,f.rhs2,f.rhs3)
f.dir <- c(f.dir1,f.dir2,f.dir3)

#solving model with Rglpk
lp_fin <- Rglpk_solve_LP(f.obj, f.con, f.dir, f.rhs, max=
    TRUE, types=types)

#----solution-----

#variables p
```

```
lp_fin$solution[1:12]
#variables q
lp_fin$solution[13:24]
#variables r
lp_fin$solution[25:36]
#variables s
lp_fin$solution[37:47]
#variables b
lp_fin$solution[48:59]
#variables c
lp_fin$solution[60:71]
#value of objective function
lp_fin$optimum
```

Numerical results

In Table 3.28 appears the result of the model. The total interests at the end of the year are equal to 1,081.47 €.

Month	p_i	q_i	r_i	s_i
1	2,500	0	0	1,000
2	0	4,000	0	500
3	0	4,000	0	0
4	2,500	0	0	1,000
5	0	4,000	0	500
6	0	4,000	0	0
7	2,500	0	0	1,000
8	0	4,000	0	500
9	0	4,000	0	0
10	3,000	0	0	500
11	0	4,000	0	0
12	3,500	0	0	–

Table 3.28: Financial plan optimizing total earnings (amounts in €)

CHAPTER 4

Bibliography

[1] Michel Berkelaar and others. *lpSolve: Interface to Lpsolve v. 5.5 to solve linear/integer programs*, 2014. URL http://CRAN.R-project.org/package=lpSolve. R package version 5.6.10.

[2] George Bernard Dantzig. Maximization of a linear function of variables subject to linear inequalities. In Richard W Cottle, editor, *The basic George B. Dantzig*, pages 24–32. Stanford University Press, 2003.

[3] Reinhard Harter, Kurt Hornik, and Stefan Theussl. *Rsymphony: Symphony in R*, 2013. URL http://CRAN.R-project.org/package=Rsymphony. R package version 0.1-17.

[4] Narendra Karmarkar. A new polynomial-time algorithm for linear programming. In *Proceedings of the sixteenth annual ACM symposium on Theory of computing*, pages 302–311. ACM, 1984.

[5] Ailsa H Land and Alison G Doig. An automatic method of solv-

ing discrete programming problems. *Econometrica: Journal of the Econometric Society*, pages 497–520, 1960.

[6] R Core Team. *R: A Language and Environment for Statistical Computing*. R Foundation for Statistical Computing, Vienna, Austria, 2014. URL http://www.R-project.org.

[7] Stefan Theussl and Kurt Hornik. *Rglpk: R/GNU Linear Programming Kit Interface*, 2013. URL http://CRAN.R-project.org/package= Rglpk. R package version 0.5-2.